CALCULATOR FUN

by David A. Adler

illustrated by Arline and Marvin Oberman

Franklin Watts
New York | London | Toronto | Sydney
1981

an easy-read ACTIVITY book

To my sister Susan

R.L. 2.4 Spache Revised Formula

Library of Congress Cataloging in Publication Data

Adler, David A.
Calculator fun.

(An easy-read activity book)
Summary: Presents simple directions for using a calculator, games to play, and some intriguing problems to solve which have surprising answers.
1. Calculating-machines—Problems, exercises, etc.—Juvenile literature. 2. Mathematical recreations—Juvenile literature. [1. Calculating machines—Problems, exercises, etc. 2. Mathematical recreations]
I. Oberman, Arline. II. Oberman, Marvin. III. Title.
IV. Series: Easy-read activity book.
QA75.A29 793.7'4 81-233
ISBN 0-531-04306-1 AACR2

Printed in the United States of America
6 5 4 3 2 1

CONTENTS

0.
C
OFF
ON
7
8
9
÷
4
5
6
×
1
2
3
−
0
.
+
=

HOW TO USE YOUR CALCULATOR

There are many different kinds of calculators. They don't all look the same, but they all have one key for each **digit**: 1 , 2 , 3 , 4 , 5 , 6 , 7 , 8 , 9 , and 0 .

They all have keys marked + , – , × , ÷ , = , . , and C .

And they all have a window, or display, where the digits you choose and the answers to problems will appear.

Your calculator should also have an "on-off" switch. Turn the switch of your calculator to "on." If your calculator is on, you should see a zero in the display.

You are ready to begin.

Addition

Press **1** and then **2**. The number **12** should appear in the calculator display.

Press **+**.

Press **3** and **9**.

Press **=**.

The number **51** should appear in your calculator display. With your calculator you have added **12** and **39**. **12 + 39 = 51.**

Now press **C**. **C** means "clear." When you press **C** your calculator "erases" the last problem. It is then ready for a new one. Not all calculators are the same. On some, you will have to press **C** twice to "erase" the last problem.

Subtraction

Press **4**, **5**, and **6**.

Press **−**.

Press **1**, **7**, and **9**.

Press **=**.

The number **277** should appear in the display. With your calculator you have subtracted **179** from **456**. **456 − 179 = 277**.

Press **C** to clear the calculator.

Multiplication

Press **3** and **8**.

Press **×**.

Press **1** and **4**.

Press **=**.

Now, **532** should appear in the display. With your calculator you have multiplied **38** by **14**. **38 × 14 = 532**.

Press **C** to clear the calculator.

Division

Press **1** , **2** , **7** , and **3** .

Press **÷** .

Press **1** and **9** .

Press **=** .

The number **67** should appear in the display. With your calculator you have divided **1273** by **19**. **1273 ÷ 19 = 67.** $19\overline{)1273.}$ with **67** written above.

Press **C** to clear the calculator.

With your calculator you have added, subtracted, multiplied, and divided.

Here are some more problems you can do for practice. Remember to clear the calculator after each problem and to turn the "on-off" switch to "off" when you have finished using the calculator.

(a) **458 + 265** =
(b) **3987 + 898** =
(c) **986 − 99** =
(d) **332 − 258** =
(e) **45 × 96** =
(f) **29 × 53** =
(g) **4272 ÷ 12** =
(h) **2176 ÷ 32** =

Answers on page 32.

STAY ON THE TRAIN!

Start with the engine of one of the "trains" shown here. On your calculator, press the keys shown on the side of the train. Press the keys in order, and stay on the same train until the last car. Press only the keys shown. When you get off the train and press the = key, a zero will appear in the calculator display.

9 + 8 + 7 + 6 +
AMO
12 × 6 × 3 − 18
2 + 3 × 12 ÷
56 − 44 × 22
4 × 6 × 5 × 2
14 × 3 ÷ 7 × 15

5 ÷ 7 ÷ 5 − 1 =
9 ÷ 11 − 2 =
15 × 2 − 8 =
8 − 33 =
8 ÷ 6 − 5 =
10 ÷ 25 − 4 =

ODD, ISN'T IT?

Here are some statements about odd and even numbers. Use your calculator and check to see if they are true.

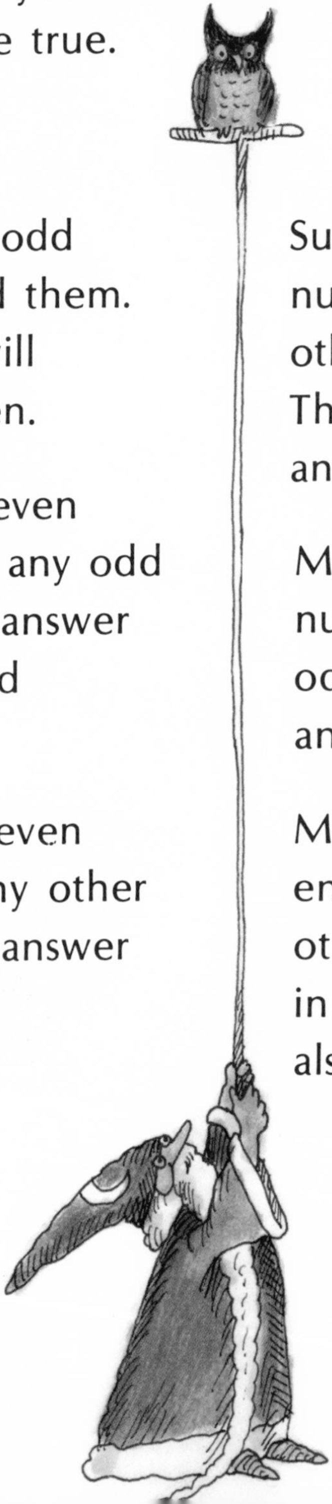

Pick any two odd numbers. Add them. The answer will always be even.

Subtract any even number from any odd number. The answer will be an odd number.

Multiply any even number by any other number. The answer will be even.

Subtract any odd number from any other odd number. The answer will be an even number.

Multiply any odd number by any other odd number. The answer will be odd.

Multiply any number ending in **5** by any other number ending in **5**. The answer will also end in **5**.

0.
OFF
ON
9
2

CALCULATOR RIDDLES

After each riddle is a problem. Use your calculator to solve the problem. Then turn your calculator upside down. When you do this, the number will look like letters. The letters will spell a word. Read the word the answer spells. It is the answer to the riddle.

1. The more you take out of it, the bigger it gets. It is a ________.
   463 × **8** =

2. It is always just the right length because it always just reaches the ground. It is a ________.
   458 + **479** =

3. It is hard to go up. It is easy to come down. It is really just a big bump in the ground. It is a ________.
   4567 + **3147** =

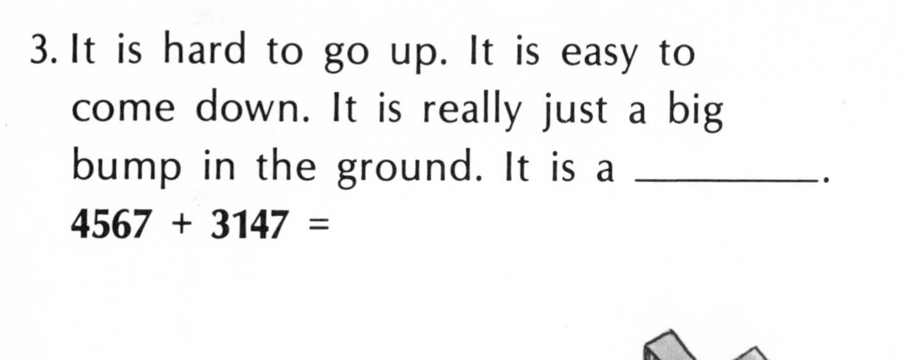

4. It has just three letters, but it is not small. What is it? ________.
   51 × 18 =

5. There is one in every minute, but an hour has none at all. It is an ________.
   2961 ÷ 987 =

6. When it is on it is never empty. When it is off it is never full. It is a ________.
   609 × 5 =

7. The more you cook it, the harder it gets. It is an ________.
   357 + 636 =

8. What comes at the beginning of every game and at the end of every song? ________.
   187 − 178 =

9. It is always buzzing around flowers.
   It is a ________.
   1319 - 981 =

10. The last thing a waiter brings you
    in a restaurant is a ________.
    3251 + 4467 =

11. What lies under the ground and
    can make a poor man rich, a clean
    shirt dirty, and a cold house warm? ________.
    142 × 5 =

12. It sits at the edge of a window but
    never falls out. It is a ________.
    10000 - 2285 =

13. It is round and long. You can take it from a tree, but you can't put it back. It is a ________.
458 + 449 =

14. It gets dirty keeping a baby clean. It is a ________.
356 + 462 =

15. What comes in every August and September but not in any other month? ________.
1000 − 995 =

16. What is in the front of every house and inside every ghost? ________.
2632 ÷ 658 =

17. It is bigger than the United States but smaller than a classroom. It is a ________.
4231 × 9 =

Answers on page 32.

H
S
e

SURPRISES

Do these problems on your calculator. The answers may surprise you.

1. A very long problem with a very short answer:
   $9 \times 8 \times 7 \times 6 \times 5 \times 4 \times 3 \times 2 \times 1 \times 0 =$

2. What happened to the 8?
   $100 \div 81 =$

3. What happened to the 9?
   $1371742 \times 9 =$

4. Blast off!

9739369 × 9 =

5. Repeaters

10 ÷ 9 =	**40 ÷ 9 =**	**70 ÷ 9 =**
20 ÷ 9 =	**50 ÷ 9 =**	**80 ÷ 9 =**
30 ÷ 9 =	**60 ÷ 9 =**	**10 ÷ 9 × 9 =**

Answers on page 32.

DOUBLING PUZZLES

A Dozen Eggs

"What's the price of a dozen eggs?" a woman asked a farmer.

"The first egg is just two pennies," the farmer answered. "The second egg is just four pennies. The third egg is eight pennies. For each egg, just double the price of the egg before it until you have paid for all twelve."

Use your calculator to find the cost of the twelfth egg.

What did the woman do?

Answers on page 32.

Working for the King

A boy had a job working for a king. After twenty-five days the king told the boy, "I'll pay whatever you ask for the work you have done as long as what you ask is reasonable."

"Give me just one small silver coin for my first day's work," the boy said. "Then just double my pay every day until I have been paid for all twenty-five days."

How much did the boy want to be paid for his last day of work?

What did the king do?

CLUE: To find out how much the boy wanted to be paid for his last day of work, press 1 on your calculator and multiply it by **2** twenty-four times.

Answers on page 32.

GAMES

The 1-to-7 Calculator Game (Two or more players)

The players take turns choosing any number from **1** to **7**. As each number is chosen, the player presses the key. Then the player presses the + key so that the numbers can be added together.

Play continues until the total is exactly **50**. The player who picks the number that brings the total to **50** wins the game. Any player bringing the total over **50** loses.

The Less-Than-10 Race (Two players)

The first player enters any six-digit number into the calculator. The second player then presses the ÷ key and then the key for any number between **2** and **9**. The players continue to take turns dividing, always choosing a number between **2** and **9.** The first player who divides and gets an answer of less than **10** wins the game.

Calculator Baseball (Two players)

The players take turns pitching and batting.

The pitcher enters any three-digit number into the calculator and then presses the × key. The batter then presses any two-digit number and presses the = key.

Whether the batter is *out* or hits a single, double, triple, or home run will depend on the digits that appear in the calculator display.

OUT	SINGLE	DOUBLE
All digits are different.	One pair of digits is the same.	Two pairs of digits are the same.

TRIPLE	HOME RUN
Three digits are the same.	Four digits are the same.

Players take turns being the batter. The batter scores zero points for an out, one point for a single, two points for a double, three points for a triple, and five points for a home run. The first player to score fifteen points wins the game.

The 500 Challenge (Two players)

The players take turns choosing a three-digit number and entering it into the calculator. After one player enters a number into the calculator, the other player must add or subtract only once and get **500** as an answer.

Each time a player gets **500** as an answer he or she scores a point. The first player to score eleven points wins the game.

The Closest Number Wins (Two or more players)

One player closes his or her eyes and enters a four-digit number into the calculator. The player writes the number on a sheet of paper, then clears the machine.

Each player then multiplies any two-digit number by any other two-digit number. The player whose answer is closest to the number written down wins the game.

ANSWERS

How to use Your Calculator

(a) **723** (b) **4885** (c) **887** (d) **74** (e) **4320**
(f) **1537** (g) **356** (h) **68.**

Calculator Riddles

(1) **3704**—hole (2) **937**—leg (3) **7714**—hill (4) **918**—big
(5) **3**—e (6) **3045**—shoe (7) **993**—egg (8) **9**—g
(9) **338**—bee (10) **7718**—bill (11) **710**—oil (12) **7715**—sill
(13) **907**—log (14) **818**—bib (15) **5**—s (16) **4**—h
(17) **38079**—globe.

Surprises

(1) **0** (2) **1.2345679** (3) **12345678.** (4) **87654321.**

(5) 1.**1111111** 2.**2222222** 3.**3333333** 4.**4444444** 5.**5555555** 6.**6666666** 7.**7777777** 8.**8888888** 9.**9999999**

Doubling Puzzles

A Dozen Eggs. The twelfth egg costs 4,096 pennies.
The woman told the farmer, "Open twelve dozen eggs and give me the first one of each."
Working for the King. The boy wanted 16,777,216 small silver coins for his last day of work. The king put him in jail.